THE AUTHORIZED BOLSHOI BALLET BOOK OF ROMEO AND JULIET

By Yuri Grigorovich and Alexander Demidov

Translated by Tim Coey
Photography by Vladimir Pchalkin
Cover: Andris Liepa and Nina Ananiashvili
Captions to color photographs by Dr. Herbert R. Axelrod

THE AUTHORIZED BOLSHOI BALLET BOOK OF

ROMEO AND JULIET

By Yuri Grigorovich and Alexander Demidov

This is a unique book. There is nothing like it in the world! Yuri Grigorovich, Ballet Master at the Bolshoi Ballet in Moscow, discloses the history of the *Romeo and Juliet* ballet and the music that Prokofiev wrote for it.

Then, with the assistance of master photographer Vladimir Pchalkin, the Bolshoi Ballet cast was photographed during live performances, in scenes which depict the complete story of this most popular ballet.

The type used in this book was set large so it could be read in dimly lit halls where *Romeo and Juliet* may be presented. The captions accompanying the photographs are ample for even a child to understand the plot and what the dancers are attempting to portray.

So here you have the only Authorized Book of the Bolshoi Ballet, magnificently illustrated with over 60 color photographs. The story of the ballet should be read to children who will be taken to see the ballet. Bring the book along so everyone will understand the meaning of the various scenes and dances.

PUBLISHED BY T.F.H. PUBLICATIONS, INC.
1 T.F.H. Plaza Third and Union Aves.
Neptune City, NJ 07753

Manufactured in the USA

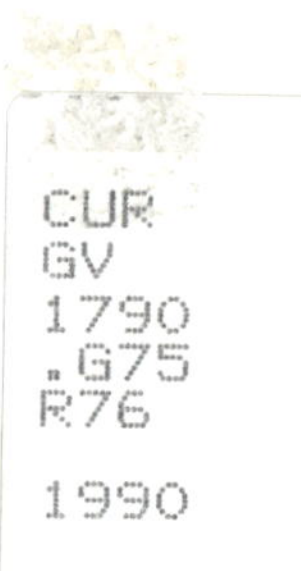

This book is available through the kind facilities of VAAP Copyright Agency of the Soviet Union.

Distributed in the UNITED STATES by T.F.H. Publications, Inc., One T.F.H. Plaza, Neptune City, NJ 07753; in CANADA to the Pet Trade by H & L Pet Supplies Inc., 27 Kingston Crescent, Kitchener, Ontario N2B 2T6; Rolf C. Hagen Ltd., 3225 Sartelon Street, Montreal 382 Quebec; in CANADA to the Book Trade by Macmillan of Canada (A Division of Canada Publishing Corporation), 164 Commander Boulevard, Agincourt, Ontario M1S 3C7; in ENGLAND by T.F.H. Publications Limited, Cliveden House/Priors Way/Bray, Maidenhead, Berkshire SL6 2HP, England; in AUSTRALIA AND THE SOUTH PACIFIC by T.F.H. (Australia) Pty. Ltd., Box 149, Brookvale 2100 N.S.W., Australia; in NEW ZEALAND by Ross Haines & Son, Ltd., 82 D Elizabeth Knox Place, Panmure, Auckland, New Zealand; in the PHILIPPINES by Bio-Research, 5 Lippay Street, San Lorenzo Village, Makati Rizal; in SOUTH AFRICA by Multipet Pty. Ltd., Box 235 New Germany, South Africa 3620. Published by T.F.H. Publications, Inc. Manufactured in the United States of America by T.F.H. Publications, Inc.

Yuri Grigorovich being honored after a performance of *Romeo and Juliet* at the Bolshoi Theatre in Moscow. The dancer, Nadezhda Pavlova, presents Grigorovich with flowers to celebrate his 60th birthday.

Nina Ananiashvili (Juliet) and Andris Liepa (Romeo) during an actual performance of the ballet in Moscow (1988).

CONTENTS

THE STORY IN PICTURES STARTS ON PAGE 65

Dr. Herbert R. Axelrod with Yuri Grigorovich in Grigorovich's apartment in Moscow after they put this book 'to bed'.

Introduction

Each major ballet production of a great composer is an enigma. Such enigmas are Tchaikovsky's *The Nutcracker* and *Swan Lake.* Prokofiev produces one with his *Romeo and Juliet.*

Before this ballet, I already had experience working with this composer's music. Indeed he is one of my favorite composers. What attracts me in his music? Probably the deeply Russian national flavor which makes his musical version of even Shakespeare's *Italian Tragedy* something Russian in essence, content and character.

Prokofiev always offers a tangy alloy of the lyrical and the grotesque, of the intimate and the dramatic. Contrasting colors, bold comparisons and deep imagery make Prokofiev theater stand a class apart. And what amazing feeling he has for dance! What a fine understanding of ballet and its esthetic particularity! What marvelous command of the orchestra! What deep feeling for truth in art!

A Prokofiev production is an undertaking which is both challenging and attractive. His world draws you in, engrossing you, and demanding an absolutely perfect ear.

With *Romeo and Juliet* I saw my goals as fully restoring the author's original score and finding a style and dance concept which would do full justice to the music. There had to be nothing external and no ideas which contradicted what Prokofiev composed. *Romeo and Juliet* is among the most integral ballets and I had to try and convey this on the stage, maintaining the neoclassic dance esthetics which grew from the traditions of Tchaikovsky, Petipa, Leo Ivanov, Glazunov and other leading figures in Russian ballet.

Yuri Grigorovich,
Moscow

Circumstances

In 1932 Sergei Prokofiev returned to his homeland, having abandoned Russia in 1918, not long after the Revolution. He had been to the USA and France where he became involved in the dance world after meeting the impresario Sergei Diaghilev. He thereby came into contact with Russian ballet choreographers including George Balanchine who then was only setting out on a great career.

In 1929 Prokofiev composed and Balanchine staged the ballet *The Prodigal Son.* In this work, composer and choreographer, each in his own way, reflected the problems of emigration. Balanchine had left Russia in 1924 and was summing up the first results in his creative life. He tried on the mask of the prodigal son, experiencing the sentimental bliss of returning. Prokofiev by that time had been abroad 11 years and was contemplating a big decision. And, as transpired later, he was evaluating a whole period in his life. He was then 38 years old and saw the world differently from the 25-year-old Balanchine.

The choreographer was giving serious thought to the idea of returning. Did he earnestly believe then in the parable of *The Prodigal Son?* Prokofiev certainly saw no shade of irony in it.

Three more years and the choreographer and composer parted. Balanchine set out to conquer America. Prokofiev settled in Moscow.

Causes

Prokofiev belonged to the Russian age of music more than any other Russian composer of the 20th century. He was a pupil of Anatoly Lyadov, a truly national composer of Russia who wrote fairy-tale miniatures, knew Russian folklore, gave unique color to his work, and breathed the poetic element of Russian life and culture.

Prokofiev also studied under Nikolai Rimsky-Korsakov who created a truly Russian national school of opera which featured a mythological, fairy-tale symbolism. Prokofiev delved painstakingly into the lessons of *The Golden Cockerel* and *Kashchei, the Immortal,* which were among Rimsky-Korsakov's later works. And the theme of Ivan the Terrible, the tsar who initiated statehood — a theme which greatly interested Rimsky-Korsakov — subsequently became major in Prokofiev's art. He wrote the music to Eisenstein's famous film. The music was later made into an oratorio, and still later it became the basis for the ballet *Ivan the Terrible* which Grigorovich presented at the Bolshoi Theater, and about which he wrote a book.

Prokofiev belonged to Russian music and culture, to its traditions and history. Thus his return home was absolutely legitimate in the philosophical sense. His short encounter with Balanchine in Paris meant, in effect, an encounter of two trends in their native culture, one purely national (which had grown up on the basis of folk music and national culture), and one international which arose from assimilation of the artistic experience of European cultures.

Yuri Grigorovich holding the painting proposed for a scene in *Romeo and Juliet*. With him are a group of dancers from the Bolshoi. Squatting in the center is Vladimir Pchalkin, the leading photographer of the dance. He took all the color photos in this book.

Sergei Prokofiev returned to Russia in 1932.

Yuri Grigorovich has an immense sense of humor. He is never known to be angry or to lose patience with a dancer.

First Ballets

Prokofiev went into composing *Romeo and Juliet* well schooled. Apart from *The Prodigal Son,* he had by 1932 the experience of four more ballets: *Ala and Lolli (The Scythian Suite), The Tale of the Buffoon who Outjested Seven Buffoons, Le Pas d'Acier* and *Above the Dnieper.*

These were all one-act productions written according to the new understanding of ballet music and the new principles discovered by Igor Stravinsky at the start of the century. The Stravinsky influence was clear in Prokofiev's early works. *Ala and Lolli* bore some traces of *The Rites of Spring* and the *Tale of the Buffoon* was in parts reminiscent of the famous *Fable of the Fox, Cock, Cat and Ram.*

And, just as was Stravinsky, Prokofiev at the early stage of his art was attracted by folklore, fairy-tale images, stylized pictures of ancient Russia, and pagan myths and legends.

But even the most severe critics noted the individuality of Prokofiev's gift, his own original manner of treating the Russian theme. Leading music historian Boris Asafiev saw in *Ala and Lolli* the continuation of the traditions of Aleksander Borodin who composed the great opera *Prince Igor.* And those traditions were national and Russian.

Comparing an early ballet by Prokofiev with *The Rites of Spring,* Boris Asafiev asserted that Stravinsky's work is "something purely exotic, an attempt by a curious, pampered, refined European to achieve unknown sensations." By contrast, *Ala and Lolli* is seen by the critic as a work born in a circle of deep national conceptions of the world and nature. Asafiev wrote: "Not for a long time (since the death of Borodin) has a voice rung out so in Russian music, singing so resonantly of the freedom and liberty of a life for the living which is not hindered by any abstractions or fear for the morrow."

In the middle Twenties, Stravinsky departed from the Russian theme. He began to elaborate on a neoclassicist style (*Apollo Musagetes, The Fairy's Kiss,* etc.) which George Balanchine would soon establish on the stage in the United States of America.

Prokofiev, after *The Tale of the Buffoon,* also left the Russian theme for awhile but did not indulge in neoclassicism. His quest lay close to that undertaken by the young Soviet ballet school, a coincidence which, at first sight, would have seemed unlikely.

In 1925 the composer wrote *Le Pas d'Acier* which was dedicated to the Soviet realities of the time. That urbanized, "industrial" ballet with its many episodes containing different characters was an original response to trends which were dominant in the theaters of Moscow and Leningrad at the time. Trends toward a new choreographic image which would reflect the post-revolution era with its cult of health, energy and physical strength, and the pathos of the construction of a new world by breaking down old moral and social views.

The Soviet ballet school in its early years rejected classical dance. It was considered that the new society needed a different kind of dance free of

Yuri Grigorovich with Simon Virsaladze in his office at the Bolshoi Theatre in Moscow. Maestro Virsaladze passed away in 1989.

dogmas and canons, dance which reflected not so much individual human feelings as the collective energy of the masses. The dance of the machines is like a poster describing in grotesque form the social types and main conflicts of the time.

Though created in far-off Paris, *Le Pas d'Acier* was just what ballet theater in the Soviet Union was looking for and surpassed the early ballet experiments of Dmitry Shostakovich, a composer molded in the new Soviet Russia. *Le Pas d'Acier* formed the basis for *The Golden Age* (1930), and *The Bolt* which were political and industrial creations. Shostakovich had gone through the Meierhold school and would later become the major symphonic philosopher of our age.

Pondering on *The Prodigal Son* and leaving its destiny to George Balanchine, Prokofiev in 1932, just before his return to his homeland, created another ballet devoted to Soviet reality. It was *Over the Dnieper.*

Le Pas d'Acier was staged by Leonid Myasin who left Russia three years before the Revolution. *Over the Dnieper* was produced by Sergei Lifar who quit Russia for France five years after the Revolution. The knowledge of all three — Prokofiev, Myasin and Lifar — as to what was going on in Russia was somewhat abstract, of course. Each had his own ideas. Each had his own path in life. Of the three, only Prokofiev was planning to return. For Myasin and Lifar (who was born in Kiev and lived near the Dnieper), all this was something exotic. Prokofiev, as history would show, thought otherwise.

Return

Returning to Russia, Prokofiev was supposed to forget his foreign days. Back then and for many years to come, official critics would recognize only his Soviet works. In 1952, for instance, one critic thus analyzed Prokofiev's early (foreign) ballet compositions: "Created during the years he was abroad, they are written in a modernist spirit and their language is to a great extent formalist. These and other flaws in his ballets allow one to speak of anti-realism features in the composer's art."

So Prokofiev in 1932 was starting from scratch. Only long after his death would the works he composed in emigration become known to Soviet audiences. Soviet people were ignorant of his operas *The Gambler* and *The Flaming Angel,* three symphonies and all the early ballets. True, Boris Asafiev did mention in *Krasnaya Gazeta* in 1928 the Paris premierè of *Le Pas d'Acier:* "the art of Sergei Prokofiev is going strong." But by the time the composer returned home, the domestic situation was beginning to change. Stalin's regime was gaining real strength and was cracking down on intellectual freedom. This did not make it easy for Prokofiev to find themes for his works.

Grigorovich teaching young dancers at the Bolshoi Ballet school in Moscow.

After Prokofiev's return to Russia, the officials ignored all of his works completed during his 14-year absence.

Yuri Grigorovich is very active in the training of his dancers at the Bolshoi Theatre in Moscow.

Romeo and Juliet and Russia

The Russian intelligentsia of the 18th century had known Shakespeare well. But particular interest in him arose in the 1820s and 1830s when romanticism appeared to free Russia of archaic classicist dogmas. By then Shakespeare's plays were appearing in translation. His works gradually became prominent in the repertoire of Russian theaters. The great poet Alexander Pushkin was a major worshipper of Shakespeare's genius.

The Russian reader and theater-goer was particularly attracted by *Richard III, Othello, King Lear* and *Hamlet.* The tragic love story of Romeo and Juliet was neglected, even though Pushkin himself wrote many words of admiration in his notes. But at that time the Russian intelligentsia was more interested in theater productions where the themes of power and just revenge prevailed.

The Russia of the 1820s and 1830s knew a growth in aspirations for freedom which was bitterly defeated. The December Mutiny of 1825 by courtiers demanding a constitution recognizing human rights, was brutally squashed, and this gave rise to "Hamlet ideas" in society. Back then, *Romeo and Juliet* seemed too provincial, and was considered not to tackle the big social issues. Perhaps that's why there were no major Russian productions of it for a long time.

Later Russian musicians would demonstrate an interest in the Shakespearean tragedy. Pyotr Tchaikovsky was one. He dreamed of an opera by that name and created various vocal pieces and a world-renowned overture connected with it. He pursued the Romeo and Juliet images throughout his career: it is reflected in works ranging from the symphonic fantasy *Francesca da Rimini* to the ballet *Swan Lake.* Tchaikovsky wrote he was alight with rapture for *Romeo and Juliet.*

He interpreted Shakespeare's tragedy in a Germanic spirit. He saw it in the purely romantic concept of "love and death." Tchaikovsky was not interested in the real Renaissance whereby fiery passions were seeking an outlet in bloody clashes with mortal results. He was not interested in the colorful, juicy lifestyle of the Shakespeare tragedy. He saw the love of the heroes as a detached, spiritual force which could not find its way out of a world of evil and violence.

Tchaikovsky produced a kind of tradition for Russian culture in interpreting the great Shakespearean tragedy. That tradition would enthrall Sergei Prokofiev more than half a century later.

Choice and Plans

Soviet drama theaters in their early years willingly presented Shakespeare's plays. A revolutionary-minded public liked the emotional characters, sharp human conflicts, and heightened tension of the plots.

As for *Romeo and Juliet,* by the Thirties the drama theaters offered a sociologized treatment of the theme which was adjusted into a clash be-

Not only does Yuri Grigorovich train his dancers to dance, but he insists upon professional acting as well.

tween the dying Middle Ages and the coming age of Renaissance. Strangely enough, the heroes were given an extremely modest place in the action.

Full-blooded Shakespearean passions could not hit the Soviet stage in the Thirties as the sphere of emotions was considered taboo and even indecent. The theater historian Boris Alpers described in the following manner how the fine actress Maria Babanova, a pupil of Vsevolod Meierhold, played the part of Juliet in the middle Thirties. Calling her a "girl with unawakened passions," the critic said "the love she played was devoid of any shade of feeling" and that "a sister loves in such a way an elder brother who is noble and manly like Romeo and so intelligent as to represent to her all that is best in humanity."

Alas, all very far from what Shakespeare had in mind. But Babanova played her Juliet at a time when Prokofiev was back in Moscow. At the time he was setting about a ballet on this theme.

Why then did Prokofiev choose *Romeo and Juliet?* The theme of love had not prevailed in any previous work by him. Indeed great passions had never particularly attracted the composer's attention. And he had not shown any inclination up to then to tackle the intricate, emotion-charged genre of a Shakespeare tragedy. The distinct lyrical tone to his music tended to bear a certain detachment.

Prokofiev naturally took into account the popularity of *Romeo and Juliet* theatrically. But could he share the banal treatment of the play which the public was being served? Could he fail to notice the artificiality of the interpretation of Shakespeare and how infinitely remote it was from the original? And could he, realizing all this, afford to go against the tide? Had he any right to ignore the situation and reckon on his personal, non-chromatic view finding a producer? And what could the composer say about the issues and problems of his time if he chose *Romeo and Juliet*? To what extent did this play match life about him? How did it fit in with the world outlook of a member of the Russian intelligentsia who was a pupil of Rimsky-Korsakov and Lyadov?

Let us not forget that the ballet would be practically the first major work by Prokofiev in the Soviet Union. It was a work on which much hinged.

The composer chose a classical subject far removed from contemporary life. And he chose to address a lamentable, tragic story, this at a time when Dmitry Shostakovich was making bold, jolly ballets.

Prokofiev elected for a smoother road in the form of a traditional love story. But he interpreted Shakespeare in a most original manner, discerning in *Romeo and Juliet* the latent themes that linked it with the modern world.

If we are to judge from Prokofiev's documents, he was an introvert. This is reflected in his music. Certainly he had his view on the theme, and his own concept for a work which took him back to the traditions of Russian romanticism and Tchaikovsky.

Esthetically, *Romeo and Juliet* represented a Prokofiev who was calling out against the neoclassicist style elaborated by Stravinsky. Reviewing 19th

Besides being an outstanding dramatic coach and ballet instructor, Yuri Grigorovich is able to dance in a highly active manner. He is probably twice as old as most of the dancers in the Bolshoi Ballet.

In Yuri Grigorovich's apartment in Moscow. Standing is the prima ballerina, Natalya Bessmertnova, who also happens to be Mrs. Grigorovich. Seated from left to right are the people responsible for this book. Lyudmila Smirnova of VAAP in Moscow (The Soviet copyright agency) . . . it was her idea to do this series of books on Grigorovich's ballet creations; Dr. Herbert R. Axelrod, the publisher of the series; and Yuri Grigorovich, the world's most prestigious ballet master and author of this book.

century Russian music, Prokofiev felt like a contemporary European artist involved in the processes of world culture.

The severe medieval rhythms describing the world of the Capulet Palace evoked an image of almighty, unlimited power. Shakespeare's caustic, cynical Mercutio is presented as a poet and musician perishing at the hands of a Tybalt who is shown as a wanton murderer on the loose and someone belonging to the upper crust of Veronese society.

Reveling in tarantella, the holiday-minded Verona contrasts strangely with a Capulet Palace which is haughtily nested in sublime oblivion and spurns the world. Verona can offer nothing against the Palace save the mischievous taunts and daredevil teasing of Mercutio. And for contempt in that world one was very crudely dealt with.

Prokofiev created more than just an image of hostility. He put his heroes up against this world of medieval cruelty and, from the outset, knowingly deprived them of their feelings. Love as a passion was seen by Prokofiev as somewhat fleeting as a flame which suddenly flares up but might go out just as quickly, and leave no memory.

The composer believed in something different. He believed in spiritual courage, in the fidelity of souls secretly dedicated by Lorenzo, the hermit monk, and joined in heavenly harmony. The world of the Capulet Palace was a godless world. Prokofiev made the marriage scene into a sacred and solemn ceremony performed in grandeur and bringing the heroes to the pinnacles of human existence. This ceremony did not really exist in the Shakespeare version but is essential to the composer's. For Prokofiev the moments where Juliet's fidelity is tested and the theme of the violence of her separation from Romeo are more important than the ardent confessions of love. When she remains true to the exiled Romeo despite the hopelessness of their future together Juliet demonstrates her inner fortitude: she cannot denounce as an enemy the person with whom she is united in heaven.

The composer intentionally isolates his heroes from the general background, constructing the world of their life as a world that is utterly individual and introversive—a world of moral and esthetic values which obstinately uphold their purity. The spiritual revolt of Romeo and Juliet is defined by Prokofiev in the novelty of his work. And we must note that he contradicts Shakespeare by giving the ballet a happy ending. The composer explained this extravagant decision with the comment that consecutive suicide by the lovers was not suitable for the genre and that the grand finale was hence justified.

The composer's friends coaxed him not to "correct" Shakespeare, though Prokofiev clearly had his own logic in deciding to do so. While not agreeing with sham optimism, he considered that the spiritual protest of the heroes would conquer. He wanted to believe in this, and was not so much contradicting Shakespeare as the realities of the day. He had to assert the moral supremacy of Romeo and Juliet over evil, violence and degraded humanity. He believed in the triumph of justice like his contemporary Mikhail

Yuri Grigorovich, the world's leading ballet master: 'Ballet for me is theatre, first and foremost.'

Sergei Prokofiev in his official Paris photograph, about 1925.

Bulgakov in whose famous novel *Master and Margarita* the heroes also received "eternal peace" and justice despite all real circumstances.

But Prokofiev was forced to leave the finale unaltered and tragic, although he had originally resolved otherwise.

Four Years

The ballet was finished by Prokofiev in 1936 but it took four more years to see the stage, though there were no direct attacks on the work. It would have been difficult to find faults in it: an inert and backward medieval world was pictured against a sparkling renaissance, and the lovers are highly romantic and majestic, devoid of all bawdiness. All corresponded to the official line, or at least could be set into such bounds. And yet the unconventionality of the composer's music and the clear newness of his concept for a comparatively long period bothered the officials and the theaters.

The ballet was unable to find a choreographer. The first choice of the composer was Fyodor Lopukhov, the choreographer of the avantgarde, one of the founders of the Soviet school of ballet, and the godfather of many choreographers, including Yuri Grigorovich on whom he had a major formative influence.

Lopukhov described his meeting with Prokofiev as follows: "At the Bolshoi Theater in the Beethoven Hall, Prokofiev was playing a work of his. I sat alongside and leafed through the scores. The music both struck me and in some measure disappointed me. I was fascinated by the composer's understanding of the heart of the Shakespearean tragedy, the essence of the verse, and the essence of each individual character. But I could not concur with the musical interpretation chosen by Prokofiev for the love of Romeo and Juliet... I expressed my doubts and wishes to Prokofiev. But he did not choose to talk to me on this matter..."

Prokofiev had his reasons for showing Lopukhov his ballet first. Lopukhov had played a big role in the formation of George Balanchine and was a choreographer who had the experience of working on Stravinsky's ballets. He was a bold artist who was seeking after, and had the same fine feeling, for the classical tradition. All this indicated that Lopukhov would be the right man to stage the ballet. But Lopukhov's memoirs show that he was unable to stomach Prokofiev's conception. So he began to argue with him about the point which bore the most principle — the interpretation of the love of Romeo and Juliet.

At that time Lopukhov was presenting mostly comic, festive ballets. He was seeking points of contact with the times and that evidently kept him from a favorable opinion on Prokofiev's work. After the reproaches of formalism which flooded down on Lopukhov in the late Twenties and early Thirties, he sought hungrily after the new, contemporary and seemingly topical themes of life. True, that only brought him a fresh portion of criticism, this time in the falsity and varnish of the times. Prokofiev's tragic world seems to have frightened the choreographer.

Rostislav Zakharov, another choreographer who was offered the first stage production of *Romeo and Juliet,* was probably frightened off by Prokofiev's style of music.

Zakharov was to stage the ballet in Leningrad at the Kirov Theater. But strangely enough, he offered the same reproaches as Lopukhov. This was extraordinary because it would have been difficult to imagine two choreographers of that time in the Soviet ballet who were more antagonistic toward each other than Lopukhov and Zakharov.

Lopukhov was a mixture of innovator and modernist. Zakharov was an apologist of socialist realism in ballet dancing and had created a heavyweight genre in choreography whereby everyday gestures replaced dancing. By the time he met Prokofiev, Zakharov had already staged *The Bakhchisarai Fountain,* a ballet based on a Pushkin ballad. This production was immediately hailed as a classic example of realist art in ballet theatricals, and gained Zakharov major recognition. Not that Zakharov was ever in oblivion, for, unlike Lopukhov, he was not seen as having suspect ideas or as engaging in dubious experiments.

Logically, Zakharov the realist seemed duty bound to stage *Romeo and Juliet* by Prokofiev, a composer who was starting out and had just written a "Soviet realist ballet." But Zakharov declined the offer. Up to then he had dealt with mediocre and primitive composers who had lacked imagination for dance and were capable merely of building dramatic mises-en-scene. So Zakharov's rejection of the ballet under the pretext that the music contained not enough passion was rational.

While the search for a choreographer continued in the Soviet Union, the ballet was staged in Czechoslovakia in a small provincial theater in Brno. That was in 1938. We know little of the production, and it can hardly have been outstanding. The average theater would have difficulty coping with such a major and complex score as *Romeo and Juliet.* But the fact remains that *Romeo and Juliet* was first staged outside the Soviet Union.

After Lopukhov and Zakharov refused, Leonid Lavrovsky appeared on the scene. It was he who finally staged the ballet, four years after it was written. That was in 1940, at the Kirov Theater, in Leningrad.

Leonid Lavrovsky

Romeo and Juliet was the ballet that really "made" Lavrovsky. Until then he had been a young hopeful, capable choreographer who had achieved several decent productions at the Leningrad Choreographic School and the Leningrad Maly Opera Theater. Prokofiev's ballet was the first real hit of Lavrovsky. We might even say that *Romeo and Juliet* was Lavrovsky's one and only hit. For many years the reputation built on that ballet made him a leading choreographer, and his prestige was undiminished by a series of clear setbacks.

Yuri Grigorovich has been recognized with the highest honors awarded to civilians in the Soviet Union.

After such a modest start, in four years after *Romeo and Juliet* the choreographer would make it to the head of the ballet company at the Bolshoi Theater. He led the Bolshoi, with a few brief intervals, for almost 20 years. His successor was Yuri Grigorovich.

The career of Lavrovsky as a dancer was comparatively short and undistinguished. He performed under his real name of Ivanov. The pseudonym "Lavrovsky" he assumed later. Being a dancer, Leonid Ivanov took part in a semi-scandalous experiment by Lopukhov. His *The Extent of the Universe* was a full-blooded attempt at dance symphony. He worked with the Young Ballet of Petrograd company which had been organized by George Balanchine.

But, becoming a choreographer, Ivanov-Lavrovsky took nothing from his teachers. Just as Zakharov, he was attracted by choreodrama whereby ballet dance mundanely represented everyday reality.

From the start of his career as a ballet-maestro, Lavrovsky took a realistic and sociological approach. He managed to thrust a social theme from a Georges Sand novel into the airy, light music of Leo Delibes' *Sylvia.* Using the music of Adam (the author of *Giselle)* and Rubenstein, he constructed a serious socio-psychological drama christened *Katerina.* And in a ballet based on Pushkin's *The Prisoner of the Caucasus,* Lavrovsky denounced autocracy and the courtier class engendered by it in society.

Just as did Zakharov, Lavrovsky firmly believed that literary drama should be the priority in ballet. For them the music was secondary to reproduction of what stage directors were doing in drama. True, unlike Zakharov, he had a feeling for dance. But, alas, that feeling was suppressed in favor of Soviet realism.

Lavrovsky was attracted to Prokofiev's ballet primarily by the literary basis to it. He indeed desired to stage a Shakespearean tragedy in the language of mime and lifelike plastics. The music would be "applied" to such an end. Lavrovsky called in the famous stage director Radlov to help arrange the action. This was not the first time Radlov had been invited to assist with ballets.

The fact of the invitation is evidence of the esthetic aspirations of a choreographer who wished to see on the stage a well-formed representation of Shakespeare's great tragedy. Lavrovsky's work with the composer took a similar vein: he was restructuring the ballet into drama in order to demonstrate social conflicts. And if Prokofiev stuck to the customary pattern of interpretation for the work of Shakespeare, Lavrovsky as a choreographer sought to check out precisely that pattern, ignoring the chamber, intimate nature of the music as well as the score's highly dramatic lyrics.

Prokofiev agreed to the compromises because he knew that otherwise his work was threatened with becoming so many sheets of music. Much was redone in the scores and they became dramatically clearer and more integral. Lavrovsky had no complaints about the treatment of the love theme. He sensed deep down the purpose of the composer and recognized his right to such a version of the theme: a change from an emotional to a spiritual interpretation.

And yet Prokofiev's scores, as a result of Lavrovsky's work on them, suffered grievous losses. The choreographer set out to re-orchestrate the scores. He motivated this decision by saying that Prokofiev's lacked the scale, color and "tonality" needed for choreography. Later, when the ballet was staged in the Bolshoi Theater, Moscow, Lavrovsky and conductor Yuri Faier further distorted the score, adding more external, pompous and showy pieces. Prokofiev had composed a work which was partly chamber music and intimate. Lavrovsky wished to make the work a major social painting. Such sham artistry was very much the order of the day.

The Performance

In January 1940, *Romeo and Juliet* was presented to the public. Lavrovsky had the courage to stage the show in a complex situation. The musicians refused to play Prokofiev and the players made poor jokes about the music. There was clear, unadulterated sabotage at hand. But Lavrovsky must be credited for persevering. The choreographer felt deep down that this was a work of genius which would eventually be a success.

He overcame the problems of the music in masterful fashion. The genre of choreographic drama asserted by *The Bakhchisarai Fountain* determined for Lavrovsky the way to interpret the score. With a minimum of dance, the accent passed over to dramatic expression which enabled the actors to express themselves in full measure. This made maximum demands on the players.

And Lavrovsky's luck held. He had excellent actors for the premierè of *Romeo and Juliet.*

The Players

The first performers of Prokofiev's ballet were actors who were young and at their peak. The oldest player was Andrei Lopukhov at 42, and this was not so old considering the character parts he played. His role was Mercutio.

Romeo was played by Konstantin Sergeev. At 30, he enchanted the spectators with his dance. Fyodor Lopukhov said of him: "In the person of Sergeev, we have a great dancer of a lyrical and romantic kind. Today he is the best 'tenor' in Soviet ballet."

The critics called Sergeev the "poet of male dance" for his exquisite well-shaped figure. He had a most elegant manner of dancing which combined smoothness and daring. M.Mikhailov, an older colleague at the theater, described his dancing as follows: "Softness and springiness enabled him, without visible effort, to make high leaps and land any jump beautifully, without any noise."

Sergeev was just made for the part. Each of his performances as Romeo drew praise for his princely style.

When he took up the part of Romeo, Sergeev already was rich in experience. He danced all the leading roles in the Kirov Theater's repertoire and starred in all the new contemporary ballets.

Lavrovsky allocated to Robert Gerbek the role of Tybalt. Gerbek was an exponent of grotesque, and possessed temperament and talent which were out of the ordinary.

Andrei Lopukhov (Mercutio in *Romeo and Juliet)* was the younger brother of Fyodor Lopukhov. He was one of the greatest character dancers in the history of Russian ballet. He wrote a first textbook for the choreographic discipline of character dance. Though he died prematurely at 49, he trained a whole generation. Among his pupils was the young Yuri Grigorovich.

Hence the stage production of *Romeo and Juliet* brought together first-class dancers. And the jewel in the crown was the marvelous Galina Ulanova.

Galina Ulanova and Mikhael Gabovich as Romeo and Juliet. Photo courtesy of Alex Young.

Galina Ulanova. Photo courtesy of Alex Young.

Galina Ulanova. Photo courtesy of Alex Young.

Galina Ulanova as Juliet

Galina Ulanova blossomed late — in the Thirties and Forties. It was 16 years before the first Bolshoi Theater foreign tours when she made the role of Juliet her own.

She was Sylphide in Fokine's *Chopiniana,* Odette in *Swan Lake,* and Giselle in Adam's great ballet. Those starring roles made her a name as a ballerina with immense ability in the romantic repertoire.

But this oneness of roles probably did not content Ulanova. In her first decade in ballet, before she played Juliet, she danced all the classical parts which fitted the nature of her gift, but was also respondent to the quests of those choreographers who professed the principles of ballet drama. She was not perturbed by the scant dance in such productions and the predominant role of mime to the detriment of ballet and choreographic expression. The ballerina liked the psychologically intricate characters of the heroines in the dramatic ballets. The vital, human content which they bore was an attraction, in spite of the stage forms in which they were realized.

As a result, when Ulanova moved from Leningrad to the Bolshoi Theater in Moscow, she refused the old repertoire parts with the exception of *Giselle* and *Chopiniana.*

At the time this caused an uproar amid the critics. Many refused to accept this step by Ulanova, viewing it as betrayal of the principles of the school she came from. Such critics said that, since Ulanova had been trained by the illustrious Agrippina Vaganova who advocated classical

Galina Ulanova. Two views.

dance, she should have developed the canons of academic classics in her art and should not contradict such canons. But Ulanova was an artist of the new type which the critics, with their concern for the problems of thoroughbred ballet, could not understand.

Ulanova sensed the times and always aspired to be contemporary. She found the boundaries of the old abstract classics a burden. She asserted her own stage style, that of psychological ballet. She understood ballet as theater which was unlike the conventional drama theater merely in that it had its own expressive devices. These features of Ulanova's talent were what made her Juliet such a roaring success. No ballerina other than Ulanova, even given the same external attributes, could have captured so fully the depth of the great lover. Nobody else could have so felt the actuality of the work. It is no coincidence that nobody since has known the success which Ulanova enjoyed as Juliet in Lavrovsky's production. And there have been many eminent performers of the part. Ulanova's Juliet was contemporary. A tragic page from history was given life and breath by the ballerina.

Ulanova had tried herself in choreodrama before *Romeo and Juliet* by playing Maria in *The Bakhchisarai Fountain.* That image was the preparation for Juliet. It was a first sketch of a later creation of greatness. Maria, in Ulanova's interpretation, embodied spiritual non-acceptance of violence and protest against such. A European captive in an Eastern harem, Maria was rendered by Ulanova with immense depth of feeling as demonstrating staunchness of spirit. This feature is consolidated by indigenous national properties.

Vera Krasovskaya, the ballet historian, said Ulanova in the role of Maria had conveyed the strength of "Russian female endurance." This pure soulfulness in Ulanova's work prepared her for the role in Prokofiev's ballet, though by the irony of fate the ballerina was originally an opponent of the work. That opinion quickly swung round, because Prokofiev's music perfectly suited the gift of Ulanova. The great actor Mikhoels, who founded the Jewish Theater, called her divine. "Ulanova makes my soul ache," he wrote.

Writer Aleksei Tolstoy also noted Ulanova's brilliance, naming her a "real goddess." And Fyodor Lopukhov, who had known personally the art of Anna Pavlova and all the great Russian ballerinas from the 1890s on, said: "the significance of Ulanova's creativity... extends beyond the confines of ballet theater."

And here's what Lopukhov's junior contemporary, Yuri Grigorovich had to say: "When after the School I went to the Kirov Theater, Ulanova was already in Moscow. But at that period she still came back frequently for performances in Leningrad. It is hard to explain, but when Ulanova danced at the theater, there was always a special atmosphere in the audience. Her uniqueness is impossible to express in words. It was as if some magic light fell on everyone around. When Ulanova came on tour, I played in some of the performances. I was the young man and Nuraly in *The Bakhchisarai*

Fountain, and the jester in *Romeo and Juliet.* The company felt exceptional responsibility when dancing with Ulanova. There was a festive atmosphere in the theater."

But, to get back to 1940, there was one more statement of opinion which was probably more significant: "She is the genius of Russian ballet, its elusive soul, its inspired poetry... In all Ulanova's creations you sense her sharp, piercing mind."

Those words belong to Sergei Prokofiev. In Ulanova he saw more than merely a polished performer of the part of Juliet, but an artist who was at one with him, and who sensed and expressed the deep purpose of his work.

The Première and the Critics

There were various opinions about the ballet. The controversy only really subsided when, after the Moscow première, it was awarded the Stalin Prize, which marked supreme approval. It was a rash person who would argue against anything thus recognized. And so for many years Lavrovsky's stage version was above criticism. The Stalin Prize virtually canonized it.

But before that event, there had been plenty of disagreement. There was broad discussion of it in Leningrad, but such material was subsequently well hidden in the archives.

The debate was heated, and opinions were mostly very critical. The people with more foresight saw the danger of Lavrovsky's version. His was a trend which took the dance out of ballet. While not demanding that he turn right back, Lavrovsky's opponents feared that the likes of *Romeo and Juliet* could bring dance down in culture, replacing choreography with graphic, illustrative forms. The opponents were not too outspoken, for all their apprehensions.

Indeed the success of *The Bakhchisarai Fountain* and *Romeo and Juliet* asserted the new choreodrama genre as the only true path in ballet. And this signified neglect for the historic ballets from the times of Petipa, Leo Ivanov, Mikhail Fokine and Sergei Gorsky.

The public naturally were thoroughly confused. The new choreodrama was attractive since it was original and expressed ideas clearly. The tempestuous, though talented, experiments of the 1920s were said to be elitist and artificial as compared with the new ballet trend. The public received well the first choreodramas, and the fine actors must have had a lot to do with that. But then the public were not asked why they liked these ballets. Before them the Theater was opening up — bright, vivid and full of effects. Theater where fine actors were expressing themselves in images could be understood by all.

Lavrovsky's opponents complained about the production. And the composer was attacked by leading critics too. The major music expert Ivan Sollertinsky, an advocate of progressive forms who never got involved with official groups, sided with Lavrovsky when reviewing the premierè of *Romeo and Juliet.* He said Lavrovsky had been innovative in "patching up weak musical conceptions." Sollertinsky was delighted with the monumental, specific decor by P. Wiliams, praised the production as true to the era and the origin, and complained that Prokofiev's heroes had proved far too abstract, with the local color unconvincing.

The critic reproached the composer for a romanticist treatment of the love theme and castigated him for not giving a sufficiently realistic image of Verona. For a man with a fine knowledge of music, Sollertinsky adopted a Stasov-like stance. (Stasov was a famous 19th-century realist art critic who was a bugbear for Tchaikovsky). Just as Stasov had failed to find sufficient "local color" and reality in Tchaikovsky's overture-fantasy on the theme of *Romeo and Juliet,* so Sollertinsky was impeding the ballet version.

It is intriguing that Lavrovsky's proponents and opponents all in their own way dodged the music, playing down the main event: an eminent ballet score was born.

Prokofiev himself had mixed feelings about the premiere. He was on the whole pleased, but noted that the success, in his opinion, could have been greater if the theater had followed the music more precisely. In those fleeting words, it is not difficult for us today to imagine the bitterness and disenchantment Prokofiev must have felt.

Moscow

In 1941 Galina Ulanova began work at the Bolshoi Theater. Leningrad's ballet fans were shocked, particularly since her partnership with Konstantin Sergeev was thus broken.

"The soft, soundless flights of Sergeev, and the airiness of Ulanova's dance were poetry, pure music," wrote ballerina Tatiana Vecheslova about the pair. "They were made in art for each other... but alas! In the very prime of their talent, circumstance has ruthlessly disrupted a couple unique in plasticity and spiritual beauty. It is difficult to fathom what a truly immense loss our art has suffered in this parting." The last remark echoed Mikhail Mikhailov, Vecheslova's colleague at the Kirov Theater.

All the best talents were then draining to the Bolshoi Theater. Leningrad, the city of the Revolution, was yielding its leading positions in cultural life to the city of Victory, Moscow, which was preparing to celebrate its 800th Anniversary in 1947.

The fates of the first performers of *Romeo and Juliet* turned out different. Death came to Andrei Lopukhov. Robert Gerbek continued his career in art until 1960. And Konstantin Sergeev, finding a new partner in the virtuous Natalia Dudinskaya with her fine technique, launched into choreography and undertook an administrative career. Neither brought him particular success or fame. The magic duo with Ulanova would never happen again.

Ulanova had quite another destiny. At the Bolshoi Theater, where Leonid Lavrovsky would soon head the ballet company, and where Rostislav Zakharov was constantly working on stage, she found new roles for herself while retaining her favorite parts of Maria and Giselle.

In 1946 Lavrovsky's Shakespeare ballet appeared at the Bolshoi — practically in the same form as in Leningrad. True, there was a lot more brass in the orchestra. But the production was more or less a carbon copy.

Lavrovsky was again fortunate with performers. The main players were exceptional. The sharp and expressive character dancer Sergei Koren performed Mercutio. Alexei Yermolaev was Tybalt. Yermolaev was among the reformers of the male dance style in the Thirties, a dancer of mighty temperament, bright color, spontaneous talent. Romeo was played by Mikhail Gabovich, a charming, elegant fellow of handsome proportions and a soft, lyrical gift.

The Second World War, and the hardship endured by the nation, altered Ulanova's performance of Juliet, and the whole performance. The ballet had been born in a different social context which had all but muffled the original conception by carrying it into a different plane. And the famous dash by Juliet across the whole stage (described by Vera Krasovskaya as "The fates of many women here arose") was interpreted in a new way in Moscow. Back in 1940 one could sense in the dash by Ulanova the real lot of women and people in love violently and unjustly separated from their loved ones. Ulanova gave a big social meaning to the scene where Juliet, downcast in pleas not to give her away to the hateful Paris, rushes in a last hope to the cell of Father Laurence. In the Moscow version that scene was construed in the light of the war years, soldiers' widows and victims of occupation. And Ulanova remained true to her inborn sense of the time.

Prokofiev's Score

For a good twenty years after the Moscow premiere, Lavrovsky's show remained the one and only interpretation of Sergei Prokofiev's ballet. Other stage productions in other cities were based on the Lavrovsky version. Any thought of a fresh view seemed blasphemous, and there was nobody in Moscow or Leningrad willing to run the risk. It wasn't until 1967 and Novosibirsk that a new version came out, and it was immediately attacked by Lavrovsky's supporters and the advocates of choreodrama.

Those were times when there was a sharp clash between original and old directions in ballet theatricals. The masters of the Thirties and Forties represented the old school. The new school consisted of artists who had emerged after the 20th Party Congress which announced "de-Stalinization." Grigorovich led the new generation.

The confrontation was fierce. The younger people were calling for a return to traditional Russian choreography, seeking to re-evaluate the negative outlook on the experiments of the Twenties. They demanded a broader range of themes in ballet art and a changeover from pompous, external monumental illustration to a quest for truth and a search into human soul and character. The advocates of choreodrama said any criticism of choreodrama was an attack on socialist realism in general. They identified their genre with the artistic medium per se.

Rostislav Zakharov spoke in such fashion at that time. He asserted that choreodrama was the method of socialist realism. This "method" actually resulted in productions devoid of content. Though the style was seen as sacrosanct and jealously guarded, it eventually split at the seams. One of the important landmarks leading up to this was the new production of *Romeo and Juliet* by young choreographer Oleg Vinogradov, who now runs the Kirov Ballet.

Vinogradov was, as might have been expected, accused of formalism and distortion of Prokofiev's work. Some people conveniently forgot that Lavrovsky had been the one who departed from the Prokofiev scores.

The fine, original production which Vinogradov engineered was still no major threat to the holy of holies. It happened far away — in a provincial city with a provincial cast. But a beginning had been made.

The Fate of Lavrovsky's Production

When Ulanova quit the stage, Lavrovsky's production endured solely due to past glory. Raisa Struchkova was a charming, touching Juliet, and Marina Kondratieva gave a lyrical, soft rendering. Nikolai Fadeechev was a worthy Romeo. Ekaterina Maximova and Vladimir Vasiliev made a fresh imprint of their own on the ballet, as did Natalya Bessmertnova and Mikhail Lavrovsky. Maya Plisetskaya produced an extremely expressive treatment for the role. But, for all that, the show was no longer in the big time. The same thing happened in Leningrad where the Lavrovsky version was maintained just as impeccably by talented newcomers. The attraction of the show waned and died.

As they say in the theater, the show flopped. There were still ballet stars to do the production justice but the ballet was just not actual any more. Lavrovsky's work lived as a relic, as an artistic monument from a past age which was hopelessly out of touch with the current times. Prokofiev's masterpiece was crying out for a new, contemporary rendition.

Grigorovich and Prokofiev

From the early Seventies there was talk that Grigorovich was ready to present *Romeo and Juliet* anew. Rumor said the venue would be Leningrad. The idea was suggested by Irina Kolpakova, a ballerina who had made a career after performing in Grigorovich's early productions in Leningrad. But the idea was not to be accomplished just then. There was a mixed bag of reasons.

Grigorovich was by then up to his eyes in work at the Bolshoi Theater. But, apart from that, there could be no real talk at that time of reviewing the former Lavrovsky version. The idea would remain merely a good intention for the future. Neither Moscow nor Leningrad was ready to give up their relic.

Chance intervened. After a success at the Paris Grand Opera with *Ivan*

Yuri Grigorovich in the Bolshoi Theatre in Moscow.

the Terrible, Grigorovich was asked to produce there *Romeo and Juliet* according to the Prokofiev score. *Ivan the Terrible* was a Prokofiev ballet which included music composed for an Eisenstein film. The only difference was that Grigorovich first staged it at the Bolshoi Theater (1975). This time the production would originally be shown abroad.

Grigorovich accepted the offer and in 1978 the premiere of *Romeo and Juliet* took place in Paris at the Grand Opera. Thereafter the question of a Moscow presentation was natural, a question of principle for Grigorovich, a controversial issue. Prokofiev's work had proven a crossroads in the history of Soviet ballet dancing.

Prokofiev's *Romeo and Juliet* was associated with the height of choreography in the Thirties and Forties. Another ballet by the same composer — *The Story of the Stone Flower* — had been staged at the Bolshoi by Lavrovsky in 1954 and was such a failure it signalled a crisis for a whole era, whereby it no longer offered any artistic prospects. But precisely with that Prokofiev work, Grigorovich launched a career in choreography three years later. And where the illustrious Lavrovsky had failed, the upcoming Grigorovich accomplished a major victory. Shortening the title to *The Stone Flower,* Grigorovich put on his version in Leningrad in 1957, marking a new phase in Soviet ballet.

Taking up work on *Romeo and Juliet,* Grigorovich naturally recalled the arguments of the past. And he remembered how he, a young man then, was called to lead the Bolshoi Ballet's company as the replacement for Leonid Lavrovsky. A new production, by virtue of all these historical circumstances, had a special meaning and significance. The correctness of the artistic method of the choreographer was being tested. His record would have been incomplete had he neglected one of the best scores of the 20th Century.

Paris

In Paris the first stars in *Romeo and Juliet* were Dominique Kahlfouni (Juliet), Michael Denard (Romeo) and Jean Guizerix (Tybalt). They had all worked with Grigorovich on *Ivan the Terrible.* The Bolshoi's Natalya Bessmertnova and Alexander Bogatyrev rehearsed the roles simultaneously and danced in one of the premiere performances.

The Paris version comprised two acts. Significant reductions were made to the score. The action unfolded on a virtually empty square, during a traditional Italian carnival. The secret characters in masks, now representing a Capulet, now a Montague, opened the show and were involved in all the action. The love scenes contrast greatly with this rowdy kaleidoscope. Later, in Moscow, Grigorovich changed the general presentation while leaving the treatment of the main theme intact. The heroes of the Paris performance were images of ideal lovers; they were a romantic couple in a pure form.

The Grand Opera performance was a clear success with the public. But the critics were dissatisfied, condemning the choreographic concept and fastidiously attempting to examine it isolated from the music. The critics examined the performance minus Prokofiev which was absolutely wrong when dealing with a choreographer of Grigorovich's rank. The critics thirsted for a more lively, contemporary approach and ignored the score and the character of the music. Prokofiev's 'return' to Paris therefore was not without controversy. After the Grand Opera premiere, Grigorovich made ready to stage *Romeo and Juliet* in Moscow. And that involved further difficulties. Some actors in the company suddenly spoke out for the Lavrovsky version which by then was a dead item in the theater's repertoire.

The argument went beyond the bounds of creative discussion. There was the play of acting interests typical of any company. A compromise resulted: the new version would coexist with the old.

This decision was no favor for the Lavrovsky production. The ardent proponents of the old ballet were not so keen about actually appearing in it. The new stage version was produced in 1979. The old could not find a cast.

The year 1979 saw a fundamentally new interpretation of Sergei Prokofiev's score. There was original treatment for the music and movement with a bias toward romantic traditions.

Yuri Grigorovich and Romanticism

The tragedy here is born in the spirit of the music. It does not seek psy-

Yuri Grigorovich.

chological motivations. The dance is action given impetus by a tragic thrust toward death.

By its very essence, music is a tragic art. Thereby music most fully accomplishes itself, as a demonic beginning to the world that bears a power which is unquestionable and defies definition. The dance born of the spirit of the music holds the same fatal mission. Thus all the romantic artists understood dance. That thought found representation in the ballet *Giselle,* a masterpiece which possibly knows no equal in the history of choreography.

The tragedy of Giselle is innate in terms of the character of the dance which engenders the tragedies and all manner of romantic heroines. Giselle just has to dance. Through dance she spares herself insanity, but love brings her to that anyway. And then her dancing, glowing with night light, comes to the realm of death, returning via sacrifice to its essence of salvation. Via dance Giselle not merely frees her loved one from the revenge of the shadows, but elevates herself to true beauty, to the love of an artist, and hence to genuine, heightened creativity. The romantics have declined the concepts of love and art in tragic conflict. In such ballets as *Sylphide* and *Giselle,* the subject is merely a condition and background for a battle of other forces, since the tragic content of such works consists in the very dance which forms its own dramatic collision.

Yuri Grigorovich created a romantic ballet. The origins of his version of *Romeo and Juliet* lie in the masterpieces of old works of the romantic era. We must also consider the experience accumulated over subsequent eras which have added their own significance to romantic themes. Thus in the stage production we have the mottos of *Sleeping Beauty,* the ingenious idea of Petipa which has often been used by Yuri Grigorovich. Thus, too, the element of stylization which exists in the choreography and conducting of this production and was partly born of romantic reminiscences of Fokine and Anna Pavlova.

Grigorovich's perception of the romantic tradition is impeccable, genuine and alive. That he uses this tradition is only natural for an artist who elaborated romantic collision in earlier works and often turned to romantic choreographic devices. His conception of the world always had a romantic kernel which is expressed even in the epic-heroic character of the performances.

Juliet was a favorite image in romantic art, and no real artist who had anything to do with romantic esthetics passed by that image. Juliet has been honored by Musse, and Berlioz who created an ingenious symphony after the Shakespearean tragedy. For Hoffmann, the author of *The Nutcracker,* the image of Julia became a fetish pursued throughout a lifetime. It found reflection in various literary fantasies and influenced real, live situations. Love for young Julia Mark, which to some extent formed Hoffmann as a writer, was inspired by the image of Shakespeare's great heroine. The very sound of the name Julia contains the boundless realms of immortality which envelop lovers spurned by life. Thus the romantics treated Shakespeare's tragedy, seeing it as a story of love and death. It means triumph for a Novalisian idea (named so after the German poet Novalis, 1772-1801, the most extreme and most gifted romanticist) whereby only death spells the true wedding night for the lovers.

Search for a Leading Lady

Grigorovich had a romantic interpretation for the role of Juliet and for the tragedy as a whole. The path to that image was marked out in his very first productions. He looked for an ideal heroine whose highest quality would be inexplicable femininity, and an ability for devoted, utterly selfless love. In Grigorovich's early ballets the hero was usually caught between two heroines who reflected the conflict in his soul: the pole of sentiment and fatal passion, and that of ideal adoration engendered by a courageous heart. This kind of apposition was typical of romanticism. Just take the stories of Hoffmann. Dual female roles appeared in many romantic ballets too. The finest example is Odette and Odile in *Swan Lake.* Such incredible duality of romantic outlook always evoked a special image of life. Perturbation and conflict of passions keynote the plot.

The dual images disappeared from Grigorovich's work as he produced ballets which were no longer purely romantic but epic-heroic tragedy. In *Spartacus* the images of the two heroines could not be parallel and linked solely with the life of the main hero. Phrygia and Aegina lived in different worlds — the former in the world of gladiators and slaves, the latter in patriarchal Rome — and the duality came from social rather than romantic distinctions. But the theme of romance lived on in Grigorovich's ballets. It appeared in *Ivan the Terrible* where an epic structure does not prevent the choreographer from expressing a personal, romantic view of history which was based on a romantic interpretation of Shakespeare's villain dramas. The romantic principle was also present in the ballet *The Hangar* which had a contemporary theme. But essentially this was a lyrical story which was intimate and full of romantic passions and sentiments.

The romantic aspirations of the choreographer were expressed in these shows chiefly through the images of the main heroine: Anastasia in *Ivan the Terrible* and Valentina in *The Hangar.* The quest for the ideal heroine culminated in *Spartacus.* Then along came the ideal image of a girlfriend, a companion in life, whereby Grigorovich has continued the images of his early leading roles for ballerinas, though on a different level. The heroine knows no antagonism. She alone dwells in the soul of the hero in whose "demonic" visions there no longer can be room for others. She creates that closed peaceful world which is ever threatened by external forces and which eventually perishes before them.

Grigorovich substantially reassessed his famed romantic view in the works which came directly before *Romeo and Juliet.* If in his early productions love was an unattainable dream which lay on a road of torment, now love became an independent pole of conflict which was involved in a clash with the forces of history and nature. This love was a union, a love both ideal and doomed, a fragile, unprotected human sentiment. Safeguarding it is the supreme obligation of a person — even after the death of one of the lovers.

Natalya Bessmertnova

Grigorovich had worked with Natalya Bessmertnova on this image of an ideal, romantic heroine. Bessmertnova was a purely romantic dancer who had been a revelation in *Giselle* and could be said to have gone down in the stage history of this production.

Bessmertnova never danced Undina or Sylphide. Apart from Giselle, she performed only the parts of Odette-Odile. The century could not be seen as entirely romantic, that is, if one considers the style and technique in which the role is cast. Only the character stipulated by the music, which reflects a definitely romantic world outlook, can be said to be romantic.

But the image of Undina, which engendered all the dream heroines of time past, was omnipresent in Bessmertnova's work, and in her every creation. Bessmertnova embodied the especially feminine character on the boundary between real and imaginary, and combined in herself flesh and infinite qualities. You could never find the dividing line. Hence the fantastic brightness of the images of the ballerina, the dream-like aspects, their illusory nature and the tragic incompatibility with the real situation on which Grigorovich built the conflict in each of her roles. Only in its own closed, intimate world can this spirit for some fleeting moment acquire reality and the fullness of earthly happiness.

Like the naiads of mythology, Bessmertnova's heroines are incapable of blending into the life around them. And the glitter of magic Sylphide-like wings which fall away on contact with a hostile world, always casts a spell in her dancing which by itself bears a tragic content. Hence the sensation of the tragic. In this sense, Bessmertnova's dancing is invariably romantic since it is all based on the conflict of the airiness and the cruel forces of the Earth. The main feature of her dancing is upward flight.

Decorated by a particular theme, this dance bears a portent by itself. Anastasia in Bessmertnova's rendering is reminiscent of the daughter of an ignorant nobleman who had the fortune to become king. Coming from such a background, Anastasia (Bessmertnova) appears as a specter at someone else's wedding, a vision of a person mutilated by fear, a spirit which suddenly acquires flesh. But these spirits paid dearly for their terrestrial happiness. And the death of Anastasia is no mere case of treacherous poisoning. She departs to her own world, one open and known only to her.

Precisely in this way Bessmertnova performs the death scene. And such is the choreographer's concept. The torment of the heroine before death is replaced by peace. She dies like the Snow Maiden (the Russian version of Undina) in the rays of the sun, sacrificially receiving the anger of nature.

Yuri Grigorovich and his wife, Natalya Bessmertnova.

From Epos to Lyrics

The image of Juliet was the artistic culmination of many years of joint endeavor by Grigorovich and Bessmertnova. It was refinement of the themes discovered by the ballerina in Giselle and by the choreographer in a whole series of ballets.

Grigorovich in *Romeo and Juliet* largely departed from the epic ballets that had attracted him in the recent past. The hero themes which were so prominent in his earlier works were muffled in this work. That required a reorientation in structure and stylistic form.

Take the duet dances. In *Romeo and Juliet* they are culminating. Previously Grigorovich had favored monologues and first turned to duets in *Ivan the Terrible.* Duet dances came to reflect ideological advances in his world. In *Romeo and Juliet* the duets set the tone and express the philosophy. They are major dance subjects accompanied by the female corps-de-ballet who personify a world of love and hope. This device, which originated in *The Sleeping Beauty* and was developed in *Giselle,* gives the duets the significance of culminating episodes in the drama. These are no mere intimate confessions or love dialogues but full-blooded dance scenes.

The duet dances in *Romeo and Juliet* can even be seen as a separate drama describing an introverted and isolated world of a love which bears tragedy in its heart and soul and is doomed to perish in the classic argument between beauty and time. That thought is central in Shakespeare's sonnets. In Sonnet CXVI he writes:

Love's not Time's Fool, though rosy lips and cheeks
Within his bending sickle's compass come;
Love alters not with his brief hours and weeks,
But bears it out even to the edge of doom.

Grigorovich aspires to embody this frail, lyrical, poetic love. He seeks such a content in every dance rather than direct narrative of the action. For example, Grigorovich does not over-organize the early scenes where Verona awakens, the serfs fight, and a clash of clans ensues. He prefers to bring events smoothly to their culmination in the general scuffle. As far as I know all other productions followed the literary letter of the plot. Other choreographers considered they must go into all the everyday details.

Grigorovich thought differently. The initial scenes were, for him, an exposition of characters rather than events. An exposition of different levels of dance (as in symphonies), of contrasting and original dance sequences in which the movement as it developed defined events and evoked the tragedy. The dance is no narrative here. The dance shows the world of Romeo, Mercutio, Tybalt in a series of plastic scenes which reveal the main ideas behind the production: a world of love aloof from reality (Romeo), a world of wild urbane festivities (Mercutio and the buffoonery) , and a world of demonic hate (Tybalt). Each character commands a group from the corps-de-ballet.

This was a new departure for Grigorovich. A historical element whereby some tale which was narrated had been a feature of past works by him. Just take *Spartacus* where each scene was a major set piece, and at the same time bore its conflict and a culmination which provided an impulse for further action.

Certainly, in that ballet the choreographer kept to symphonic principles, just as in *The Stone Flower, Legend of Love* and *The Nutcracker.* But *Romeo and Juliet* was a further development in his concept of ballet drama.

The dance elements in *Romeo and Juliet* are free of all sense of narration. Without being fragmentary, Grigorovich is classical in outlook. There are two dance flows, two different outlooks which, like themes in a symphony, suddenly clash as is inevitable.

Similar dramatic constructions are to be found—in Act Two of *Giselle,* in *The Sleeping Beauty* and in a scene in *Swan Lake.* But classically this style was not used throughout a ballet. So Grigorovich was totally innovative.

Tybalt

A tragedy in music is formed against the background of the merry city of Verona, which is at the same time a leading character. The crowd of people running about free of care constantly clash with the element of evil embodied in Tybalt and his retinue. The dancing is light, airy and fast as opposed to "heavy" dance which seems fixed down by gravity. All Tybalt's jétès are like hysterical cries and have characteristic hand movements which seem to prime his body in readiness for flight. The plasticity of Tybalt and the corps-de-ballet group following him, conceal an uncompromising resolution. They advance as the guardians of enmity, the forces of darkness and a hate which is vented against the whole world.

Berlioz said Shakespeare's Tybalt was a "genie of anger and revenge." The first statement Tybalt makes in the play is one of hate for the world. "I hate the world!" he exclaims, and only later does he name Montague as a specific target.

Tybalt's hate bears its own brand of drama. It can never be quenched since it knows no bounds. Grigorovich puts a demonic halo on Tybalt. He is the proud knight of evil, the spirit of hate and hostility, a figure somewhat abstracted from reality, the dark cloud over a jolly Verona. The plastic outline of the role contains intonations of Spanish dance. Such are the specific pirouettes of Tybalt, certain poses and positions of his hands which paint an image of a foul matador who is elegantly adorned and takes life as an endless battle.

In the scene where Tybalt is killed, the choreographer's metaphor finds a new sense. The matador seems to have changed places with his victim. Blinded and poisoned with hate, he is like a wounded, cornered beast. As though fatally injured, he heads for Romeo who is standing with cape; Tybalt is ready to wipe out everything on his last path.

Tybalt's world is the male world. There are no women in his retinue which exists in terrible isolation from them even in the ballroom scenes. The ball in Grigorovich's production is a ritual ceremony, a fantastic parade of live mannequins. The image is one of fatal strength, a ball of spectres. They emerge before Romeo front-stage, and suddenly in the background, and move inexorably upon the hero.

This world has its own laws and traditions and its life determines the especial mechanism of passions, rituals and habits which do not accept any life other than themselves. Indifferent to time, this world keeps a tight hold on an elitist isolation. The hatred lies in disdain for life. The enmity is only a means to preserve the aristocratic world which hides the spirit of ancient knighthood — magnificent and cruel at the one time, fine but doomed to inevitable death which explains the aspiration for revenge.

The images in this world will appear to Juliet in the scene where she imbibes the drink prepared for her by Father Laurence. As though her mind is wandering before death, there is a march before Juliet of ladies and knights, the spectres of the familial crypt reminding her of a neglected duty before the laws of that world in which she did not wish to be happy. The march cuts diagonal paths across the stage, moving like a funeral procession, arousing in the mind of the heroine an image of her own burial, the path to eternal rest, to the haven of past generations whose laws and traditions are sacred to the world surrounding Juliet.

This episode is multi-faceted and is prompted by the famous monologue of Juliet. Even death, real or illusory, does not hold promise of salvation from the world of shadows, and this thought drives her to despair. There is no way out of her situation.

The world of the Capulet family is one of showy ritual, a world strangely numbed after the death of Tybalt. The wail of griefstricken women and calls for revenge by shocked men acquire a special significance in Grigorovich's rendering. The stage is lit in red, and the colors of the costumes also range in red, the crimson shades predominating. The atmosphere is one of universal trouble, fatal disaster, of a whole world crumbling, veiled in fire amid which there are moans, cries and screams betokening the agony of further deaths.

Tybalt is a spirit of that world. In the street fight episode, he dashes about the square as though summoning more and more crowds of people. He incites the battle and is virtually a participant himself. He suddenly grows up out of a heap of interwoven bodies, above a crowd beaten into one lump and seemingly drunk with blood. He orders them to continue the battle. He seems to be battling with everybody — with the other side and his own side, making no distinction in the heat of the moment and hating life, an emotion which brings him into conflict with Mercutio.

Mercutio

If Tybalt's world is one doomed and finished, Mercutio's is presented by the choreographer in bright, vital tones. Mercutio's image in the ballet is linked with the theme of the city: the bright and prospering Verona comes through in the classical nature of the dance which gives a poetic atmosphere. Prokofiev's music has no ethnographic elements to be adhered to. Grigorovich, in complete harmony with the score, elects for character-popular dances as classical dances, in the tissue of which individual folk elements are organically weaved. The local color is sensed in the fabric of these dance compositions.

There are also Italian features (tarantella movements) but they are not the general style. Nor are the grotesque, eccentric elements which exist for this character. The main thing is the classical beauty of any hair-raising show of plasticity, eccentric escapade, and the most paradoxical dance combination. Mercutio's dancing is most virtuous and full of different intonations. Nevertheless it always is classified as harmonious and beautiful, which befits a role first played by the memorable Mikhail Tzivin.

Mercutio is accorded a special place in the drama of the production. He does not interest the choreographer as a hero in some Renaissance tragedy or a definite true-to-life figure representing the progressive, humanist forces of that era. That is the usual interpretation of Mercutio, but Grigorovich envisages him as connected with a romantic tradition and the romantic reminiscences of Carlo Gozzi's commedia dell'arte. It's an estheticized look at a past culture thanks to which Mercutio is a lyrical character who reflects the author's voice which nearly has a nostalgic feeling.

Mercutio here is a lyrical buffoon, the embodiment of the spirit of life as abstracted from the specific realities of the topic. He contains a spirit in art which is full of fantasy and noble mischief and which thirsts for adventure whatever the danger. Unlike that of Tybalt, Mercutio's world does not exclude the female form. The life of his world is multi-form and multi-colored. In it reigns a dance element which offsets the closed eliteness of Tybalt.

Mercutio is accompanied by masks: characters dressed in the suits of the Italian comedy and outwardly reminiscent of Callo. Only externally, for these characters lack Callo's cruelty or violent obsession. Their sphere is jocularity rather than cold mockery. Humor, and not cynical irony. Light-hearted buffoonery kept in good measure.

Grigorovich's masks are mainly musicians: they accompany Mercutio and play as he dances, forming an orchestra of street minstrels. Their image sets off the tragedy, adding a new element to the atmosphere. At the ball, the masks constantly break into the prim festivities to add some luster and life in the background. But they bear no particular significance. The masks here are no ironic commentators. They do not parody the main characters but remind us of another aspect of life that is open and joyous rather than formal and ritual.

Just one mise-en-scene involving the masks bears special significance. That is the moment Tybalt comes out to the ball for his dance, which is magnificent and frightful, full of mighty emotions. The masks then surround him as though involuntarily and he ends up in a ring of characters mockingly accompanying him and enveloping the rants of his fierce hatred in fun and jokes. Tragedy opens up for one moment a frontier beyond which comes a cosmic area heralding the end of the world. The masks are an image of free-style dance and the image of Mercutio bears the same portent in the production.

Mercutio all the while invites Romeo to go on a journey — a long walk leading to death. This dance is partially built diagonally as it needs plenty of space. The mask in his hands is no more than an accessory, a piece of clothing; it is not hiding anything. Even mortally wounded, he, supported by the comedians, continues to move, unable to stop. His dancing lingers a moment as though filmed in slow motion. He floats smoothly up in the hands of his assistants and slumps down into the splits. And a further step up to the sky, a run at the air broken off by death. Down on earth, surrounded by kneeling companions, he makes a final effort to fly up but his body is fixed to the earth.

Mercutio dies in a paradoxical mise-en-scene. Leaning on his shoulders and chest, he falls still in a vertical position as though his legs, those of a dancer and comedian die last at the moment of some unthinkable dance which seems to go on after his heart has stopped and his soul has departed him.

There will be none of the animation as when Mercutio at the ball plays out an episode of death. But that death was not from a dagger but from the dance. And dance movement revived him. That was a romantic parody of death but strangely heralded his actual death. Mercutio's path in the production is that of a true musician who has lived out his own life in dance, and runs counter to a world soaked in hatred.

Love

The tragedy of Romeo and Juliet develops in the sphere of dreams that confound reality. Though Romeo at times takes part in Mercutio's jokes, Juliet behaves at the ball as a girl of the world should. Nevertheless the choreographer keeps his lovers apart from real life. The spirit of life (Mercutio) and the spirit of hatred (Tybalt) define different poles of tragedy which lead to conflict and fatally influence the world of love of the main characters which is one aloof from earthly passions.

Their links with reality are illusory. They endure reality only under compulsion. In this temporary and ethereal world they are waiting to go to a new life eternal and elevated. In the first scene of the ballet which presents Romeo, the choreographer declares the romantic isolation of the world of love from everyday living, even where that is clear of mundane objects.

The curtain rises and we see a stage covered in darkness: a cold and unbounded lifeless desert, the starless gloom of the barrenness of a world not yet created. Then a ray of light illuminates a young man sitting in the square. The light seems to summon him out of the ignominy of that immense universe, seems to call him its hero and chosen son, its Romeo. Magical spirits fly out to the hero from all parts of the square, girls in pink tutus with sleeves which flutter like wings in the wind, creating expectations. The world is aflood with new colors, the first rays of dawn, the first signs of awakening.

The theme of life arousing is synonymous with the birth of love. The sun rising over a dreary medieval Verona is a symbol of feelings being born. The dance of the main hero, his first monologue, gives the impression of night flight, of an as yet unclear confusion of a soul full of the sensations that come before dawn.

The visions — dreams or illusions of the hero — dart in front of him, surrounding him and leaping. The girls emerge from all corners of the stage, responding to every movement of the chosen one. Airy grands jétès dominate the hero's movement. These are appeals to the night. The corps-de-ballet play visions which form a pink-fiery train behind the dreamer-boy. The train represents wings born of a yearning for love which persists until he meets Juliet. Romeo romantically rejects reality for quest after a mysterious ideal which will bring him death.

This yearning of the hero is the highlight of the first scene in the ballet which is close to the Shakespearean version of Romeo wandering by night and meeting the crimson light of dawn at one with nature. It is this solitude and melancholia which give Romeo his romantic aura and allow him to ask: "What is love?" Thus the theme is set for the entire play:

"Love is a smoke rais'd with the fume of sighs;
Being purg'd, a fire sparkling in a lover's eyes;
Being vex'd, a sea nourish'd with lovers' tears:
What is it else? a madness most discreet,
A choking gall and a preserving sweet."

Visions

The corps-de-ballet play visions which appear in all the love scenes and accompany the hero and heroine from birth to death. But these visions have a freer role in Grigorovich's ballet version, in which they acquire some independence. They are woven into the choreography of Juliet's part as envoys from a fairytale world.

After the prologue the visions reappear at the first kiss of the lovers. That dawn of love is blessed by a world of spirits. Then there is the famous confession of love by Romeo and Juliet in the Capulet garden by night where the dancing of the corps-de-ballet gains impulses as the lovers' feelings intensify. Expressive dashes give the atmosphere of that sacred moment, that Shakespearean exaltation of love.

At the moment of marriage, the visions virtually become real girls. With flowers in hand, they form a kind of wedding cortege uniting the fates of the lovers. They do not hasten to help Juliet at the most terrible moment of solitude when, after betrothal to Paris, the heroine loses faith in a happy outcome. But the images of love return the hope. In thoughts of Romeo she seeks salvation. Father Laurence appears offering the magic potion. And Romeo swims through Juliet's mind. The duet of the lovers is a possible finale. A joyful good ending full of harmony. After this most successful piece of choreography, Juliet accepts the potion, assured that her love will be accomplished.

The corps-de-ballet is involved when the lovers perish. They die on a bare, open stage, and in the distance, behind a veil, we see the contours of a vault. By the tomb of Juliet spirits of love stoop. First they stand in a semicircle and then, inclining, they seem to cover it with their hands, now blessing this abode of love, now defending it.

As life departs the lovers, the visions of the corps-de-ballet, like fading flowers, droop lower and lower until hands alone remain.

The flower metaphor is a most favorite device from the neo-romantic age. Just recall Fokine's *Vision of a Rose* as performed by Anna Pavlova where the main theme is the ephemeral nature of a doomed beauty. Flowers, butterflies and dragonflies were the images of frailty and transient happiness that the great ballerina professed.

The metaphor of the flower is that of a defenseless world of visions which is unable to withstand cruelty. Hence the classical images of Giselle, Sylphide and Undina. And the Juliet created by Yuri Grigorovich and Natalya Bessmertnova belongs to that category.

The Spirit of the Music

The Moscow production of *Romeo and Juliet* shows just how sensitive an ear Grigorovich has for the music. The net fragments in the score round off the ballet, giving it the unity it once lacked. The second act used to lose out in terms of expression since the fates of the lovers were off stage, while the deaths of Tybalt and Mercutio gained the main accent. And despite the good effect of the latter episodes, they could not compensate for the total absence of the main tragic theme.

Grigorovich introduced a fresh episode in the Capulet garden where Juliet shows her feelings after the night rendezvous with Romeo. And thanks to this, the three main scenes in Act Two are now centered around the love theme. This is more in keeping with Shakespeare.

Act Three also gains a new interpretation and is probably the most refined part of Grigorovich's production. Restoring the dance of the jesters (part of Prokofiev's original score) into the scene with the tragic wedding morning of Paris, the choreographer creates an episode with great emotional intensity. There are dramatic contrasts. Paris is presented as a noble, sincere person, who is somewhat cool in mannerism and devoid of the passion of a character which Shakespeare had actively competing for Juliet.

The Paris played by Mikhail Gaborovich is passive and inert, a chance guest at a tragic festival. His role is cast in a gallant, knightly detachment.

In Lavrovsky's edition, Paris had a satirical tone. He was a poseur, a vain person constantly looking in a mirror carried about everywhere by a page. Other productions also had Paris as a trite character: the choreographers did not want to give Paris the serious qualities he bears in Shakespeare's play. Grigorovich restored the nobility of the image, and thereby he gave it back its drama.

Paris dances in the accompaniment of young men and women who have

Simon Virsaladze, the Bolshoi artist and set designer, one of Grigorovich's closest friends and colleagues. He died in Georgia, USSR, in 1989.

come to congratulate the bride. The picture is full of bright colors and bears no trace of the looming tragedy. The Oriental dancers and musicians are designed to entertain the young people. Their dancing is used to heighten the action by adding an element of eccentricity. The heady compositions are connected with Paris's personality: he represents a world of unfulfilled dreams, and his ordinariness keys us up for a grand finale.

The dance of the jesters ends the suite and transfers it into grotesque, tragic farce. The dancing is a travesty in which fate gibes. The jesters wave their fingers menacingly at the guests, clasp their stomachs and shake with laughter as if knowing that the bride is doomed and they have come to a wake rather than a wedding. The action here reaches a climax and happens at an illusory border between happiness and disaster, farce and tragedy. At the end of the suite, Paris and the whole worldly assembly are caught up in the rhythm of the buffoons' movement, merge with them and fly along with them toward the couch of the bride. And at this moment of exuberance, the Capulets appear, bearing the body of Juliet in their arms...

Simon Virsaladze

The romantic concept of the production is based on a duality. The stage is fully available for dance as in *The Stone Flower* or *Spartacus.* Only in the distance, behind a veil, are there elements of real life: the Gothic outline of Verona, the moonlit Capulet garden, Laurence's sunlit cell, and the luxury Capulet residence with its sculptures and sparkling columns.

Virsaladze's stage decor is in keeping with the mythological, eternal interpretation of the plot. The picturesque backdrops are like illustrations in an antique book. The space on the stage gives real life to a legend that dates back to the early years A.D. but lives on infinitely thanks to the immortality of human soul and mind.

Virsaladze finds perfect harmony. The costumes are just right for the dance which arises from the music. The choreographic symphony passes over into a symphony of art which is emotionally captivating. This applies particularly to Act Three with its unsettled, almost autumnal atmosphere of sadness. There is a sense of chill overhanging hot Verona, a feeling of nature fading and dying (thus the dying flowers) : the love story departs into eternity, into a torpid calm.

Bessmertnova as Juliet

Juliet was for Natalya Bessmertnova the final word. The ballerina presents an ideal Juliet who is never mundane. By analogy with the music, one can say that Bessmertnova's performance lies in the domain of the absolute, that it is exceptionally artistic.

Bessmertnova's Juliet is never literary. Her heroine is just made for ballet dancing. Only in the first episode in Juliet's chamber are there specific details: the image of a merry, lively girl. But at the ball there is a changeover and the reborn Juliet is already a mythical, romantic heroine.

True, the choreographer leaves her in some everyday scenes: with the childnurse (Act Two), and mourning for Tybalt. But these episodes rather prompt the ballerina to go on to show the genuine, poetic side of her character. Tragedy dwells in the very dance of Bessmertnova, in its romantic upward tendency to meet with fate. The culmination of the role comes in the final duet dance of the lovers which forms a grand finale of love and death.

In this finale, Grigorovich departs from Shakespeare to choose Da Porto's novella version where Juliet wakes before Romeo is poisoned. Grigorovich thereby achieves a particularly resonant tragic finale which highlights the lovers' destiny of death.

Death finds Romeo at the tragic apotheosis of love. He reels back from the tomb on seeing Juliet rouse. She appears to him as a vision of death. The lovers have to part, for the poison has been imbibed and not even their affinity can change events. The first sign of death is when Romeo lifts Juliet so that the figure of a cross is formed. Just as love seems saved, it is severed. This is the tragic metaphor of the scene where Romeo's body slowly slips down, as though interred by the body of the one for love of whom he has given up his life. Bessmertnova at this point represents the epitome of tragedy. The final mises-en-scene of the duet are convulsive embraces as they try to wrest themselves clear of the cold numbing throes of death.

Juliet dies on her knees as though departing into a dream. Her death is quiet for the most terrible thing is now behind. Juliet, in Bessmertnova's rendering, meets death with a sensation of inner harmony, with knowledge of performing a duty before Nature.

Finale

Grigorovich finds a truly romantic finale for the ballet. There is no traditional relenting of the hostile parties. There are no pompous oaths nor any other effects which would seem inappropriate at the grave of dead children.

When the lovers die, they remain prostrate center stage and slowly out come all the characters of the ballet (except Mercutio, Tybalt and Paris). They gradually fill in the space and conceal the dead bodies of the lovers. In the hands of each character dimly burns a candle and in this vast sea of people neither the Montagues nor the Capulets can be distinguished: all are equal before Lady Death and subjugated to Her will and power.

And slowly over this sea of candles, the prone lovers are borne up by the corps-de-ballet. They lie side by side and lack any semblance of death. All Juliet's contours are turned toward Romeo and they give the impression of being merely asleep. These are lovers who have found eternal peace and an eternal repose of love which Father Laurence blesses on his knees, praying to the skies as though marrying those known as Romeo and Juliet.

Then the light fades gently from the stage and only the candles twinkle in a world shrouded in darkness.

Afterword

Grigorovich created a romantic ballet—a ballet in the fullest sense of the word. He broadly and boldly interprets the traditions and laws of ballet production. More than translating Shakespeare's tragedy into the language of dance, he represents it in a brand new way. Therein lies the value of the production. It is no mere illustration; it expresses the spirit of Shakespeare's *Romeo and Juliet* in movement.

The choreography is truly harmonious from start to finish. This version of Sergei Prokofiev's ballet is a major event in the history of dance. This is ballet at its enigmatic and beautiful best.

Romeo and Juliet, as presented by Lavrovsky, knew broad acclaim due to a social content which was read into the score. Galina Ulanova guaranteed the production international fame. But that time has gone, never to return. Grigorovich's is a deep, historically objective rendition.

Grigorovich approaches the work as an esthetic and not social phenomenon. He calls on the classical traditions of dance and art and shows the romantic side first and foremost. The world of his Romeo and Juliet is not rippled by the storms of everyday life. Unlike Maurice Bejart, who presented a Berlioz ballet of the same name, there are no contemporary associations. Grigorovich intentionally goes into a world apart. The controversy which this elevated approach has evoked does not keep the ballet from retaining its popularity.

Last season Grigorovich turned once more to *Romeo and Juliet:*

"Not the first time I have remolded parts of the choreography," says he. "I at one stage rejected the visions and then, partially restoring them, reduced their part; after all there is a lot of music in it, even though it is great music. The artist has over the years changed some things in the set and costumes. Maybe the first version was too abundant. Now I want to make things more rounded and laconic without reviewing the basic concept. Well, time marches on and opens up more and more new layers in Prokofiev which require further refinement. I am sure that the ballet world has yet to fully comprehend the works of this great composer."

Two housefolds, both alike in dignity,
In fair Verona, where we lay our scene,
From ancient grudge break to new mutiny,
Where civil blood makes civil hands unclean.
From forth the fatal loins of these two foes
A pair of star-cross'd lovers take their life;
Whose misadventur'd piteous overthrows
Do with their death bury their parents' strife.

William Shakespeare

Yuri Grigorovich standing in front of the Bolshoi Theatre in Moscow (1987).

SYNOPSIS

ACT ONE

SCENE ONE — The main square of Verona. A quarrel and a fight between the Montagues and the Capulets, the two most powerful clans in Verona

SCENE TWO — The Capulet palace. The nurse hurries to Juliet, the daughter of the Capulets, to get her ready for a ball.

SCENE THREE — Romeo, the heir of the Montagues, and his friend Mercutio, disguised in masks, arrive at the ball in the Capulet house among other guests. Romeo meets Juliet and they fall in "love at first sight."

SCENE FOUR — The Capulets' garden. Juliet is out on her balcony. Romeo appears seeking for Juliet. The two young people declare their eternal love for each other.

ACT TWO

SCENE ONE — The main square of Verona. The people are having a feast. Mercutio and his friends mock Romeo, who is deeply in love.

SCENE TWO — The Capulet House. Juliet begs her nurse to take a secret letter to Romeo. Finally the nurse agrees.

SCENE THREE — Romeo receives Juliet's letter and rejoices.

SCENE FOUR — The cell of the good Friar Laurence and the secret wedding of Romeo and Juliet.

SCENE FIVE — The streets of Verona. A devastating fight between Tybalt, the nephew of the Capulets, and Mercutio breaks out. Tybalt, Verona's best swordsman, kills Mercutio. Romeo continues the duel for his dead friend and kills Tybalt.

ACT THREE

SCENE ONE — Juliet's room. The lovers' secret meeting before their painful parting. Romeo has to flee the city. Paris' proposal to Juliet.

SCENE TWO — Romeo is banished from Verona with no one to share his solitude.

SCENE THREE — Friar Laurence's cell. Laurence gives desperate Juliet a potion to make her appear dead and thus escape the wedding ceremony with Paris.

SCENE FOUR — Juliet's room. Juliet pretends to give her consent to her marriage with Paris and takes the potion.

SCENE FIVE — The Capulet house. Paris throws a ball as an interlude to the wedding ceremony. Sr. Capulet brings the news of Juliet's unexpected death.

SCENE SIX — The ancient tomb of the Capulets. Romeo's despair at what seems to him to be Juliet's death. Romeo's farewell to Juliet. Both lovers commit suicide.

EPILOGUE

The victory of eternal love over animosity and hatred between the two clans.

Verona is a medieval town in Italy, ruled by a benevolent prince. The two leading families of the town, the Montagues and the Capulets, have a long-standing family feud which erupts almost everywhere the two members meet. This causes grief to the otherwise peaceful town and the Prince finally declares severe punishments, like banishment or death, to the next trouble-makers.

Even this edict doesn't stop the manifestation of hatred which results when Romeo, heir of the Montague family, disguises himself and crashes uninvited into a private party given by Lord Capulet for his friends.

Juliet, heiress of the Capulets, sees Romeo and the two fall in 'love at first sight.' She is devastated to learn who he is and that night goes out onto her balcony to share her secret with the stars. Romeo risks his life to steal into the garden to catch a glimpse of Juliet. He hears her stellar confession and is overwhelmed. He identifies himself and the two vow their eternal love regardless of the consequences.

They are secretly married the next morning by Friar Laurence, who devoutly prays that this marriage will end the family feud.

Juliet goes home and Romeo joins his friends in the streets of Verona. They meet the fuming Tybalt, nephew of Lord Capulet, who is still angry over Romeo's intrusion into the private ball of his uncle. The argument leads to a fight during which Romeo's best friend is killed. Mercutio, the victim, is a clown who thought the fight with Tybalt was only in jest. Romeo goes mad over the sight of his fallen comrade and he grabs a sword and kills Tybalt. The Prince arrives on the scene and banishes Romeo from Verona.

Juliet's father, ignorant of the secret marriage, promises her hand to the visiting Count Paris. Juliet is in utter despair. Her husband, Romeo, is banished and she must marry Count Paris. She goes to Friar Laurence and pleads for help. The Friar prepares a drug for her which makes her appear dead, but actually she will wake up in a few days. She takes the drug immediately before her wedding and is discovered in death by her father. She is taken to the family tomb and interred.

Romeo hears about her death before he receives Friar Laurence's explanation. He goes to the tomb and sees her lifeless body. In grief he takes poison. Just as the poison begins to take effect, Juliet awakens. She holds the dying Romeo in her arms, realizing the tragedy of the situation. When he dies, she kills herself with a knife plunged into her heart.

Romeo is dreaming. It is right before dawn. His dreams are expressed in his solo dancing. He manifests a boyish, mischievous character punctuated by a subtle melancholy and youthful impetuosity. The solitude of his dance may be chilling.

The sun has risen and the communal square in the Italian city of Verona comes to life. An active group of young men and women throng around Mercutio, since he was the first to appear on the scene. Mercutio is a witty and clever but slightly devious man. These characteristics seem to make him appealing to the group around him.

More and more people throng to the square and, with flashing smiles and obvious gaiety, they begin a cheerful tarantella dance.

The fast tarantella is interrupted by the sudden appearance of Tybalt, the nephew of Sr. Capulet. The Capulet family is one of the two most distinguished of Veronese families. They are constantly embroiled with the Montagues, Verona's second most prestigious name. Tybalt is the best swordsman in Verona. He is accompanied by his three best friends.

Mercutio is the embodiment of the gay, merry Veronese spirit, while Tybalt is an evil genius. . .a demonic knight.

Tybalt is an aggressive warrior. He is egotistical about his strength and dueling ability. He cannot live without fights and he implores young men to join him in a life of mischief and adventure.

Spoiling for a fight, Tybalt strokes the flames of hatred between the Capulets and the Montagues. Romeo is the leader of the Montagues and supporters of the two sides begin their struggle. He tries to reconcile the two sides, but to no avail. The devastating fight breaks out.

The Crown Prince of Verona appears in the midst of the turmoil, with the fighting at its peak. He is not at all amused and only wants peace and prosperity in Verona.

The Prince imperiously stops the fighting between the irreconcilable families. He wants neither fighting nor bloodshed in sunny Verona.

The Prince threatens everyone. Lay down your arms and stop fighting or you will be banished from Verona and sent into exile.

The Capulets are preparing for a ball. Paris, a young nobleman, is preparing to propose marriage to Juliet. They will meet for the first time. Juliet's nurse is busy with preparations for the ball.

Juliet, the shining star of the house of Capulets, is in pleasurable anticipation of the ball.

Juliet with her parents. They have informed their daughter about Paris' intentions and are preparing her for the meeting with her future fiance.

The ball in the house of the Capulets begins. Juliet dances with Paris as a token of their first meeting.

In a circle of her contemporaries, the light-hearted Juliet dances with Paris.

Mercutio has induced his pals, including Romeo, to come to the ball in masks so they wouldn't be recognized. Surrounded by musicians, Mercutio entertains the guests in a friendly, joyous manner.

Mercutio is to the dance what Tybalt is to the sword. Everybody admires his dancing ability and his comic actions entertain everyone.

Sr. Capulet, with his Lady, begins a solemn dance to indicate they are of an older generation.

The knight's dance gains momentum as they dance around the elderly Lady Capulet.

Right: Tybalt is not satisfied with being the best swordsman in Verona; he also wants to be the best dancer, so he demonstrates his dancing skills.

Even though he is masked, Romeo doesn't take his eyes off Juliet. She, too, senses something extraordinary. This, their first meeting, will go down in history as the immortalized "Love at first sight."

Romeo and Juliet leave the guests in the main hall. They are alone. Their love ruptures the serene peace of the moment and Romeo removes his mask. They then tell each other of their great love for one another. It is a scene that everyone in the world can appreciate.

Tybalt enters at the moment of the declaration of love. Romeo succeeds in putting back his mask, but Tybalt is very suspicious.

Right: The ball is over. . .but Romeo cannot leave Juliet. He finds his way to the garden. Juliet can't sleep either and she goes onto the balcony. She sees Romeo and descends to him in the garden.

ACT TWO

The Second Act opens in the main square of Verona. The street is filled with cheerful youths dancing the traditional tarantella.

Left: Romeo and Juliet in the garden where, in the deep silence of the night, they declare their eternal loyalty. The dance is a great adagio.

Mercutio tries everything to get Romeo into a happier mood. . .but all his attempts are in vain. Romeo is a Montague while Juliet is a Capulet. What lies in store for them?

Left: Mercutio is the center of attraction in almost every merry-making event taking place in Verona. He is little more than the town clown.

The nurse finds Romeo and gives him Juliet's note. Romeo is overjoyed.

Left: Juliet asks her nurse to take a letter to Romeo. The nurse is frightened and horrified. . .she refuses. . .then, understanding the great love, finally agrees.

Mercutio's friends tease the horrified, frightened nurse. They torment her until the nurse becomes enraged.

The frail Juliet firmly rejects Paris, a husband chosen by her parents. Her father is outraged. Her mother is also firm. Juliet must marry Paris.

Juliet is inconsolable. Her parents have deprived her of any possible happiness. She must become Paris' wife as such were the customs of those days.

She runs to the cell of Friar Laurence. Can he help?

The good Friar gives her a potion which will make her appear dead. Then, when she wakes up, Romeo will return from Mantua for her.

Juliet takes Friar Laurence's potion. She begins dreaming.

In this dream, Romeo appears. He is again with her.

The house of Capulets prepares for the wedding. Paris has already arrived and with him are his friends.

Before the wedding takes place, Paris decides to entertain the guests. He has brought with him some special Oriental dancers. Their emotional dance serves as a prelude to the wedding ceremony for the two noble families.

Another of Paris' entertainments are the jesters. Both male and female dancers, their heads jingling with bells, perform dances to amuse the guests.

At the height of the entertainment, Sr. Capulet appears, holding the lifeless body of his daughter. The guests are shocked. Her nurse prostrates herself. Paris, who is guilty without guilt, rushes to Juliet.

Learning of Juliet's death, Romeo rushes from Mantua to Verona. Seeing his beloved in the Capulets' burial vault, Romeo takes poison to join Juliet in heaven. Then Juliet awakes. She doesn't know that Romeo took poison and rushes to him.

The death adagio, as staged by Yuri Grigorovich, is brilliant. Juliet doesn't know about Romeo's suicide and is frightened by his erratic behavior.

Romeo dies of the poison. Can Juliet live without him? She shares with Romeo the superb closeness that only true love can bring. It can only be compared with Adam giving a rib to Eve. Juliet takes the knife from near Romeo's rib and runs it through her own heart.

The last kiss of the eternal beloved. The heavens open to receive their forbidden love.

This is the final scene of the ballet. All Verona turns out to pay homage to Romeo and Juliet. The Capulets and Montagues put aside their hatred. The immortal beloved are raised heavenward.

In Verona, Italy, a bas-relief symbolizes this event, in honor of Shakespeare.

Romeo and Juliet are married the next morning in the cell of the good Friar Laurence. The marriage is a great secret and Friar Laurence hopes it will heal the breach between the Montagues and the Capulets.

After the wedding ceremony, the traditional poetic adagio is danced by the couple. The dance is filled with expressions of tenderness and love.

Juliet, in Yuri Grigorovich's interpretation, is like a romantic sylphid coming to a sinful Earth to remind people that love is the greatest and most humane feeling.

Mercutio starts the merriment once again.

Mercutio and Tybalt meet. Mercutio is from a neutral family, neither a Capulet nor a Montague, but he is Romeo's friend. That is enough for him to be Tybalt's enemy. Tybalt challenges him to a duel.

Left: Tybalt is frantic. He doesn't want dances. . .he wants duels!

Mercutio dies. . . The despicable Tybalt is triumphant. . .and the tender-hearted Romeo despairs.

Left: Romeo cannot stop the duel. Mercutio thinks this is merely a big hoax and he has no intention of a serious duel. Tybalt quickly slays Mercutio and takes pride in his repugnant victory.

Romeo becomes infuriated with Tybalt. He grabs a sword and courageously battles Tybalt. Tybalt, remember, is the best swordsman in Verona. But Romeo wins and succeeds in killing Tybalt. Romeo is immediately horrified at his deed.

Juliet and her mother despair over the death of Tybalt. Remember that Tybalt is Juliet's cousin. Tybalt's mother is Juliet's mother's sister.

Juliet's sorrow is two-fold. Not only does she lose Tybalt, her cousin, but she knows the Prince will banish Romeo from Verona.

ACT 3

Only God and the good Friar know of the wedding. The act opens with a love adagio signifying the consummation of the marriage. The scene is Juliet's bedroom.

The Prince will banish Romeo so this is their first and last night of love.

Parting is horrible and painful. Romeo and Juliet's last joys and torments of love.

The "sacred night of love," as Shakespeare put it, is over. The awakening is terrible. Paris, Juliet's fiance, arrives coincidentally with the exile of Romeo, to claim Juliet in marriage. Romeo must leave on the day that Juliet will marry Paris.